Thank You, Apple Trees

Do you like eating apples? Thank an apple tree. Farmers grow apple trees in **orchards**.

Pressing apples makes apple juice. Peeling and cooking apples turns apples into applesauce.

Thank You, Plants

by Luana K. Mitten and Mary M. Wagner

Table of Contents

ROURKE CLASSROOM RESOURCES
The path to student success

You might be surprised to find all the things we can thank plants for.

Apple trees also **provide** shade on sunny summer days.

Thank You, Corn Plant

Do you like eating corn on the cob?
Thank a corn plant.

Farmers grow corn in fields. They harvest it for corn on the cob or just the **kernels**.

Some farm animals eat corn.

We eat corn kernels too. Popping makes a hard kernel puffy.

Thank You, Cotton Plant

Do you like feeling soft cotton? Thank a cotton plant. Farmers grow cotton in fields.

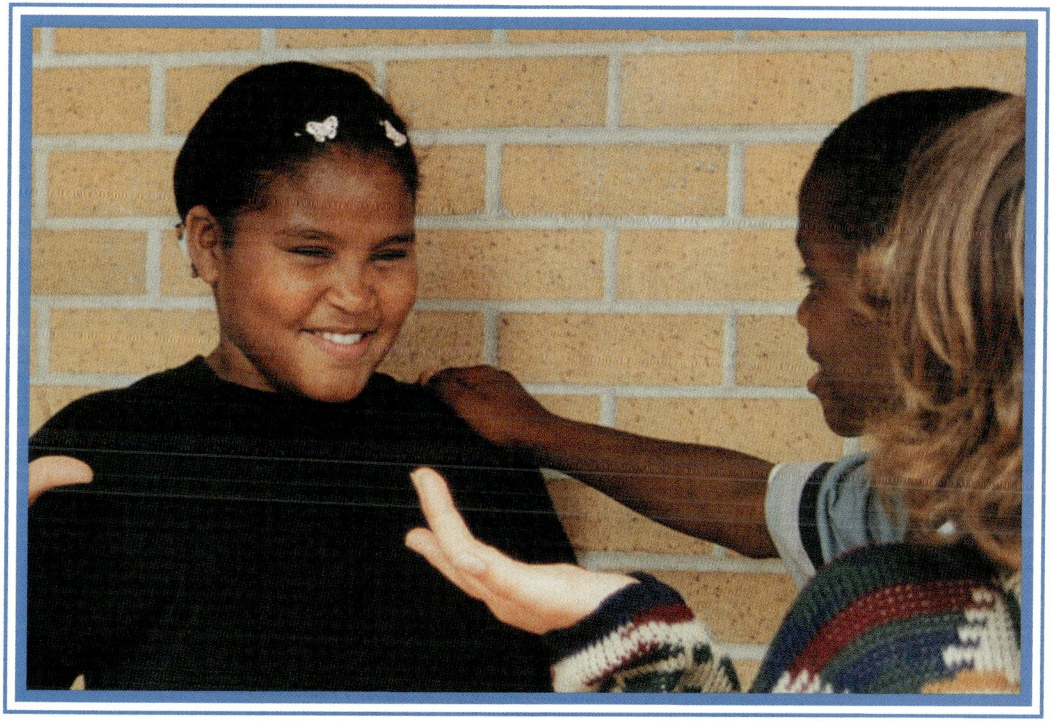

Some clothes are made from cotton. Is your T-shirt made from cotton? How can you find out?

Some writing paper has cotton in it.

Crushing cottonseed makes oil. Sometimes we fry foods in cottonseed oil.

Thank You, Rose Bush

Do you like smelling roses? Most rose farmers grow bushes in greenhouses.

Roses can be used to say "I love you!"

Thank You, Plants!

AND thank you farmers for growing apples, corn, cotton, roses and many other plants, too!